The Hadean Eon

Earth's Fiery Origins and the Birth of a Planet

Authored by
Zahid Ameer

Published by

Goodword eBooks

Copyright © 2024 Zahid Ameer

All rights reserved.

ISBN: 9798302609854

DEDICATION

"I dedicate this book to my beloved parents, whose wisdom I hold in the highest regard. Their every word of guidance has been a beacon of light, illuminating the path of my life and shaping the essence of who I am."

The Hadean Eon

Contents:

Contents:

Chapter 1: Introduction to the Hadean Eon

Chapter 2: Formation of the Solar System

Chapter 3: Earth's Fiery Youth

Chapter 4: The Formation of the Earth's Crust

Chapter 5: The Early Atmosphere and Oceans

Chapter 6: Life in the Hadean?

Chapter 7: Earth's Transformation Towards Stability

Chapter 8: Hadean Mysteries and Scientific Exploration

Chapter 9: Conclusion: The Significance of the Hadean

Glossary of Terms

Bibliography

Acknowledgments

Disclaimer:

About me

The Hadean Eon

The Hadean Eon

Chapter 1: Introduction to the Hadean Eon

The **Hadean Eon** is the first chapter in Earth's long and complex geologic history. Spanning from approximately 4.6 billion years ago, when the planet began to form, to around 4 billion years ago, this period is often viewed as the most hostile and tumultuous phase in Earth's development. The term "Hadean" is derived from "Hades," the Greek god of the underworld, a reference to the extreme conditions that characterized the planet at the time, including intense heat, volcanic activity, and constant bombardment by cosmic debris.

Defining the Hadean Eon

The **Hadean Eon** is unique because, unlike the more recent eons, it is not defined by the presence of life, tectonic activity, or easily distinguishable rock formations. Instead, it encompasses the time during which the Earth itself was forming. Modern geologists have reconstructed the events of the Hadean by studying ancient minerals, moon rocks, and meteorites, as direct evidence from this time period on Earth is extremely scarce.

The timeline of the Hadean begins with the **formation of the Solar System** from a massive cloud of dust and gas

The Hadean Eon

known as the **solar nebula**. Gravitational forces caused the material in this nebula to coalesce, forming the Sun at its center, while smaller clumps of dust and gas began to aggregate into what would eventually become planets, including Earth. Earth's formation is believed to have occurred through a process known as **accretion**, where dust particles collided and stuck together, gradually growing larger over millions of years. These planetesimals—small, solid objects that would become planets—merged, creating enough mass for gravity to shape the early Earth into a spherical form.

The Hadean Eon ended roughly 4 billion years ago, marking the transition to the **Archean Eon**, where the Earth began to cool and stabilize. However, during the Hadean, Earth was in a constant state of flux, with a molten surface and an evolving atmosphere.

Importance of the Hadean Eon

The Hadean Eon is vital for understanding how Earth developed into the planet we know today. Despite the harsh conditions, the Hadean laid the groundwork for many of the systems and structures that would support life billions of years later.

1. **Intense Heat and Volcanism**: During the Hadean, Earth was a hot and volatile world. The heat came from several sources. First, the energy released by

The Hadean Eon

constant collisions as planetesimals merged together kept the surface molten. Additionally, **radioactive decay** of isotopes, such as uranium, thorium, and potassium, generated a significant amount of heat in Earth's interior. This molten state led to the formation of what scientists call a **magma ocean**, a surface covered by molten rock. Over time, as the Earth began to cool, the outermost layer would have solidified to form the earliest version of the planet's crust, although this crust was continually being remelted by volcanic activity.

Volcanism was rampant in the Hadean as Earth's surface tried to cool, but the hot interior kept feeding magma to the surface. This volcanic activity released gases into the atmosphere, a process known as **outgassing**, which was responsible for creating Earth's early atmosphere. Unlike the oxygen-rich atmosphere we have today, the early atmosphere of the Hadean was composed of **carbon dioxide**, **water vapor**, **methane**, **ammonia**, and other gases that were toxic to modern life forms.

2. **Cosmic Collisions and the Moon's Formation**: One of the most significant events in Earth's history occurred during the Hadean Eon: the formation of the Moon. Scientists believe that around 4.5 billion years ago, a **Mars-sized body** named **Theia** collided with the early Earth. This cataclysmic impact ejected

The Hadean Eon

a large amount of material from Earth's surface into space, which eventually coalesced to form the Moon. The impact also had a profound effect on Earth's development. It may have significantly altered Earth's rotation, tilted its axis, and even accelerated its cooling by causing additional material to be ejected into space.

The **Moon's formation** was crucial for Earth's future stability. The Moon's gravitational influence helped stabilize Earth's axial tilt, which plays a vital role in regulating the planet's climate over long periods. Without the Moon, Earth's climate may have been too unstable to support the development of life.

3. **Building Blocks for Life**: Although the Hadean Eon was a period of extreme environmental conditions, it was also the time when many of the **building blocks for life** were established. The constant volcanic outgassing and the cooling of Earth's surface allowed water vapor in the atmosphere to condense, leading to the formation of Earth's first oceans. While there is still debate about how much water was present during the Hadean, it is likely that at least some oceans or seas existed by the end of the eon, providing a critical environment for the eventual development of life.

Some scientists suggest that **organic molecules**, the

precursors to life, may have formed during the Hadean. These molecules could have originated from volcanic outgassing, chemical reactions in Earth's early atmosphere, or even from **extraterrestrial sources** like comets and asteroids. It is possible that the intense energy from the sun, lightning, and volcanic activity helped drive the formation of these organic compounds, setting the stage for the development of primitive life forms in the following eons.

4. **Establishing Earth's Geologic Structure**: During the Hadean, Earth's **internal structure** was also forming. As the planet cooled, it began to differentiate into distinct layers: the **core**, the **mantle**, and the **crust**. Heavier elements, such as iron and nickel, sank toward the center to form the core, while lighter silicate minerals floated to the surface to form the early mantle and crust. This differentiation was crucial because it established the **tectonic and volcanic activity** that would continue throughout Earth's history.

The **magnetic field** of Earth may also have originated during the Hadean. The movement of liquid iron within Earth's outer core likely generated the planet's magnetic field, which plays an essential role in protecting the planet from the harmful effects of solar wind and cosmic radiation. This magnetic shield would become particularly important for the development of life in later eons, as it helped preserve Earth's atmosphere from being stripped away by solar radiation.

5. **The Role of the Late Heavy Bombardment**: The Hadean Eon also saw the **Late Heavy Bombardment (LHB)**, a period when the inner planets of the solar system—including Earth—were bombarded by a large number of asteroids and comets. This event occurred approximately 4.1 to 3.8 billion years ago, near the end of the Hadean Eon. While the LHB caused massive destruction on the surface of Earth, it may have also played a crucial role in delivering additional **volatile elements**, including water and possibly organic molecules, to the young Earth. The bombardment also shaped Earth's surface by creating craters and contributing to the planet's continued cooling.

Conclusion

The Hadean Eon is a fascinating yet mysterious time in Earth's history. It was a period of extreme heat, violent cosmic impacts, and dramatic changes that laid the foundation for the planet's future. During this time, the Earth formed, differentiated into layers, and eventually cooled enough to form a solid crust and an atmosphere. The formation of the Moon, the emergence of the first

The Hadean Eon

oceans, and the establishment of Earth's magnetic field were crucial milestones in Earth's evolution. Though no life existed during the Hadean, the processes that occurred set the stage for the origin of life in the subsequent eons, making the Hadean an essential chapter in the story of our planet's history.

Chapter 2: Formation of the Solar System

The Solar System's origins lie in a remarkable and complex process that unfolded over billions of years, beginning with a collapsing cloud of gas and dust that coalesced to form our Sun and its accompanying planets. This chapter explores the Solar Nebula Hypothesis, the formation of proto-Earth, and the violent beginnings that shaped both Earth and its Moon.

The Solar Nebula Hypothesis: The Birth of the Solar System

The Solar Nebula Hypothesis is the most widely accepted explanation for the formation of the Solar System. This theory posits that about 4.6 billion years ago, a vast cloud of gas and dust—called the solar nebula—began to collapse under its own gravity. This collapse may have been triggered by the shockwave of a nearby supernova explosion, which provided the necessary force to disturb the delicate balance of pressure and gravity within the cloud.

Stage 1: Collapse of the Nebula

The Hadean Eon

As the solar nebula collapsed, it began to spin faster due to the conservation of angular momentum. Just as a figure skater spins faster by pulling their arms inward, the collapsing cloud increased its rotational speed as it contracted. This rotation caused the cloud to flatten into a disk-like shape, with most of the material concentrating in the center where the Sun would eventually form.

The temperature in the central region of this spinning disk rose dramatically as gravitational energy was converted into heat. At the very core of this disk, the growing pressure and temperature ignited nuclear fusion reactions, marking the birth of our Sun. The remaining gas and dust in the surrounding disk continued to orbit the nascent Sun, setting the stage for the formation of planets, moons, asteroids, and comets.

Stage 2: Formation of Planetary Building Blocks

As the solar nebula cooled, microscopic dust grains began to stick together through a process called accretion. Over time, these particles clumped together to form larger and larger aggregates, ultimately growing into kilometer-sized bodies known as planetesimals. These planetesimals were the seeds of future planets.

In the inner part of the solar system, where temperatures were higher, only materials with high melting points—such as metals and silicates—could survive. This led to the

The Hadean Eon

formation of the terrestrial planets, including Earth, from rocky materials. In the outer, cooler regions of the disk, where water and other volatile compounds could condense, gas giants like Jupiter and Saturn began to form, accumulating thick atmospheres of hydrogen and helium.

Proto-Earth: The Formation of Early Earth

The process of accretion, where smaller planetesimals collided and merged, led to the creation of larger bodies over time. One of these growing bodies was proto-Earth. During the early stages of the Solar System, proto-Earth was not the serene planet we know today but a hot, molten, and highly dynamic world.

Accretion and Gravitational Forces

Proto-Earth's formation took place over tens of millions of years as it swept up smaller planetesimals in its orbit. These collisions released enormous amounts of energy, causing the young planet to remain molten for an extended period. Gravitational forces played a critical role in this process, as larger bodies attracted more and more material, increasing their mass and, in turn, their gravitational pull. This runaway growth allowed proto-Earth to rapidly gain size and become one of the dominant bodies in the inner solar system.

The Hadean Eon

As proto-Earth grew, it began to differentiate internally. Heavy elements like iron and nickel sank to the center, forming the planet's core, while lighter silicate materials rose to form the mantle and crust. This differentiation would later be crucial in developing Earth's magnetic field and tectonic activity, but during the early Hadean, Earth was a rolling sphere of molten rock.

The Role of Volatiles

Water and other volatile compounds were likely delivered to proto-Earth by icy bodies from the outer Solar System, including comets and water-rich asteroids. These materials may have been key in forming Earth's early atmosphere and, eventually, its oceans, though the timing and exact source of Earth's water remain areas of active research.

Violent Beginnings: The Giant Impact Hypothesis

In the early Solar System, collisions between planetary bodies were frequent and often cataclysmic. One of the most significant of these impacts is believed to have played a critical role in shaping Earth's future: the giant impact hypothesis, which suggests that the Moon was formed from the debris of a colossal collision between proto-Earth and a Mars-sized body known as Theia.

The Theia Collision

The Hadean Eon

Around 4.5 billion years ago, while Earth was still forming, it is thought that Theia, a protoplanet orbiting in the same region as Earth, collided with the young Earth at a glancing angle. This impact was so powerful that it caused a massive amount of material from Earth's mantle and Theia to be ejected into space. The core of Theia is believed to have merged with Earth's core, contributing to the increased size of Earth's metallic core.

The debris that was flung into orbit around Earth coalesced over time to form the Moon. This scenario explains why the Moon's composition closely resembles that of Earth's mantle. Unlike other moons in the Solar System, which often formed independently of their parent planets, the Moon is unique in that it appears to have been born from a violent collision.

Evidence Supporting the Giant Impact Hypothesis

Several lines of evidence support the giant impact hypothesis. One key piece of evidence is the similarity in isotopic composition between rocks from the Moon and Earth's mantle. This suggests that the material that formed the Moon came primarily from Earth. Additionally, the Moon has a relatively small iron core, consistent with the idea that it was formed from the outer layers of Earth and Theia, leaving the heavier iron cores to merge with Earth's core.

Moreover, computer simulations of planetary formation support the idea that such giant impacts were not only possible but likely in the chaotic early Solar System. These simulations show that planetary bodies like Earth would have been subjected to numerous collisions as they grew and cleared their orbits of smaller planetesimals.

Aftermath of the Theia Collision

The impact with Theia had profound consequences for Earth. The energy released by the collision would have melted large portions of Earth's surface, creating a global "magma ocean." Over time, the Earth began to cool, allowing solid crust to form on its surface, though volcanic activity would remain intense for hundreds of millions of years.

The formation of the Moon also played a significant role in stabilizing Earth's axial tilt, which helps maintain relatively stable seasons and climates over long periods of time. Additionally, the gravitational interaction between Earth and the Moon led to tidal forces that would have influenced the early oceans and may have played a role in the development of life.

Conclusion

The formation of the Solar System was a dynamic and often violent process. From the collapse of a giant cloud of

The Hadean Eon

gas and dust to the formation of planetesimals, proto-Earth, and the Moon, these early events set the stage for everything that followed. The Hadean Eon, Earth's earliest period, was shaped by these cosmic forces, leaving behind a planet that would eventually cool and stabilize, ready to give rise to oceans, continents, and, eventually, life. The Solar Nebula Hypothesis and the giant impact hypothesis not only help explain Earth's formation but also provide insight into the processes that govern planetary formation throughout the universe.

Chapter 3: Earth's Fiery Youth

The Hadean Eon was a period of dramatic and chaotic change for the young Earth, a time when our planet was nothing like the blue, life-filled world we know today. Instead, Earth's surface was a hellish landscape dominated by molten rock, intense volcanic activity, and a constant bombardment by space debris. This chapter explores three critical aspects of Earth's fiery youth: the "Magma Ocean" phase, the formation of the early atmosphere, and the cataclysmic bombardment by asteroids and comets during the Late Heavy Bombardment.

The "Magma Ocean": Earth's Molten Beginnings

As the Hadean began, Earth had just formed from the accretion of planetesimals—small, rocky bodies that coalesced under the force of gravity. The intense collisions between these objects generated an enormous amount of heat, so much so that Earth was, at this time, largely a molten mass. This phase, often referred to as the "Magma Ocean," describes the state of Earth's surface, which was a vast sea of molten rock.

The Hadean Eon

- **Molten Surface**: Early Earth was extremely hot, with temperatures high enough to melt most of its surface materials. During this period, the planet had little or no solid crust—just an ocean of liquid rock stretching across its surface, occasionally punctuated by rising volcanic islands that would quickly sink back into the molten sea.
- **Heat Sources**: Several factors contributed to Earth's searing heat during this time. First, the ongoing accretion process generated substantial heat as planetesimals collided with Earth. Second, the radioactive decay of isotopes like uranium, thorium, and potassium inside Earth's core produced internal heat. Finally, the giant impact that formed the Moon also created tremendous heat, further keeping Earth's surface in a molten state.
- **Differentiation of Earth's Layers**: One of the most critical processes during the Magma Ocean phase was the differentiation of Earth's layers. As Earth's molten surface gradually began to cool, heavier elements like iron and nickel sank toward the center, forming the planet's core. Lighter elements, such as silicon and oxygen, rose toward the surface to form the mantle and the earliest versions of Earth's crust. This differentiation laid the foundation for the Earth's internal structure: the core, mantle, and crust that we are familiar with today.

The Hadean Eon

- **Crust Formation**: As Earth continued to cool, the magma began to solidify, forming the first thin crust. However, this crust was likely unstable, frequently sinking back into the molten interior due to continued volcanic activity and high surface temperatures. The first truly stable continental crust wouldn't appear until much later in Earth's history, but the seeds of it were sown during this time as small patches of solid rock began to form and cool.

The Early Atmosphere: A Toxic and Hostile Environment

While the Earth's surface boiled with molten rock, its skies were equally hostile. The early atmosphere of Earth was vastly different from what we know today. Far from being breathable, the atmosphere of the Hadean Eon was toxic, thick, and hot—composed primarily of gases emitted from continuous volcanic eruptions in a process known as **outgassing**.

- **Volcanic Outgassing**: As the Earth's surface remained largely molten, active volcanoes spewed vast amounts of gas into the atmosphere. These gases included carbon dioxide (CO_2), nitrogen (N_2), water vapor (H_2O), sulfur dioxide (SO_2), methane (CH_4), and ammonia (NH_3). There was almost no

The Hadean Eon

free oxygen (O_2) in the atmosphere during the Hadean, making it completely inhospitable to life as we know it. The atmosphere was thick and oppressive, trapping heat and creating a greenhouse effect that kept Earth's surface scorching hot.

- **Formation of Water Vapor**: One of the key components of Earth's early atmosphere was water vapor, which likely came from two primary sources. First, outgassing from the Earth's interior released significant amounts of water vapor. Second, some of Earth's water may have been delivered by comets and other celestial bodies that bombarded the planet during this period. Over time, as Earth cooled, this water vapor began to condense and form the first oceans, but during the Hadean, most of it remained suspended in the thick atmosphere.
- **Lack of Oxygen**: Oxygen, the gas essential for most modern life, was notably absent in Earth's early atmosphere. This was a reducing atmosphere, dominated by hydrogen-rich gases like methane and ammonia. It wasn't until much later in Earth's history, during the Archean Eon, that oxygen began to accumulate, thanks to the advent of photosynthetic organisms. During the Hadean, however, any free oxygen would have been rapidly consumed by chemical reactions with volcanic gases or surface minerals.

The Hadean Eon

- **A Hostile Climate**: The early atmosphere trapped immense heat, creating a "runaway greenhouse effect." This meant that Earth's surface was extremely hot, with temperatures well above the boiling point of water, ensuring that no liquid water could exist on the surface. The thick, heavy clouds and intense volcanic activity would have made Earth's skies appear dark and oppressive, with frequent storms, lightning, and perhaps even acid rain due to the sulfurous gases in the atmosphere.

Bombardment by Space Debris: The Late Heavy Bombardment

As if Earth's fiery surface and toxic atmosphere weren't hostile enough, the planet was also under constant bombardment by space debris during the Hadean. This period, known as the **Late Heavy Bombardment** (LHB), occurred between approximately 4.1 and 3.8 billion years ago and was characterized by a spike in the number of asteroids and comets that pummeled the inner planets, including Earth.

- **The Origin of the Late Heavy Bombardment**: The LHB is thought to have been caused by the gravitational interactions between the outer gas giant planets—Jupiter, Saturn, Uranus, and Neptune—and

The Hadean Eon

the leftover planetesimals from the solar system's formation. As these gas giants settled into their current orbits, their gravitational pulls sent waves of asteroids and comets hurtling toward the inner solar system, leading to a period of intense bombardment.

- **Impacts on Earth**: During this time, Earth was repeatedly struck by large asteroids and comets, some of which were as large as small cities. These impacts would have caused widespread destruction on the planet's surface, vaporizing parts of the crust and possibly even re-melting large sections of the mantle. The energy released by these impacts was immense, with some estimates suggesting that the energy of each collision was equivalent to millions of nuclear bombs.
- **Consequences of the Bombardment**: The impacts from the LHB had several important effects on Earth. First, they likely played a role in shaping Earth's surface by creating massive craters and possibly disrupting early tectonic activity. Second, the impacts may have contributed to the formation of Earth's oceans. It is theorized that some of the water present on Earth today was delivered by icy comets during this bombardment. However, the intense heat generated by the impacts would have vaporized any surface water at the time, meaning that liquid water

likely couldn't persist until the bombardment ended and Earth's surface cooled further.
- **The Moon as Evidence**: Much of what we know about the Late Heavy Bombardment comes from studies of the Moon. The Moon, lacking an atmosphere or significant geological activity, has preserved a record of these ancient impacts in its surface craters. By studying lunar rocks and crater formation, scientists have been able to estimate the timing and scale of the LHB. Given that Earth and the Moon are close neighbors in space, it's reasonable to assume that Earth experienced a similar bombardment.

Conclusion: The Path to Stability

Earth's fiery youth was a time of chaos and upheaval. From its molten surface in the Magma Ocean phase to the toxic and hostile early atmosphere, and finally, the relentless impacts during the Late Heavy Bombardment, the Hadean Eon was a crucible in which our planet was shaped. Although this period was marked by extreme conditions, it was also a necessary phase in Earth's development. Over time, as the planet cooled, the crust solidified, and the atmosphere began to stabilize, Earth set the stage for the next chapter of its history—the Archean

The Hadean Eon
Eon—when life would finally begin to emerge on this once-hostile world.

Chapter 4: The Formation of the Earth's Crust

The Hadean Eon represents a period of intense transformation in Earth's history, particularly when it comes to the formation of the planet's first solid crust. After Earth's initial formation as a molten mass, the planet underwent significant cooling, leading to the solidification of its surface. This chapter explores the processes involved in the cooling and solidification of Earth's surface, the emergence of the earliest continental crust, and the evidence provided by Earth's oldest rocks, which help scientists reconstruct the conditions of the Hadean Eon.

Cooling and Solidification: How Earth's Molten Surface Began to Cool

In the earliest stages of the Hadean Eon, Earth's surface was dominated by molten rock. The young planet, still accreting from collisions with planetesimals, was heated by a combination of factors: the kinetic energy from these impacts, gravitational compression, and the radioactive decay of elements like uranium, thorium, and potassium. This intense internal heat caused Earth's surface to remain in a near-constant molten state for millions of years.

The Hadean Eon

Eventually, as the rate of cosmic impacts slowed and the planet began to lose heat to space, Earth's surface started to cool. The cooling process was complex, with different areas of the planet solidifying at different rates. Cooling occurred primarily through radiation into space, as the heat from the Earth's surface dissipated. Over time, a thin, unstable crust began to form as the molten rock cooled and solidified. This initial crust, often referred to as the **primordial crust**, would have been highly dynamic, continuously forming, breaking apart, and being recycled into the molten mantle beneath due to the intense volcanic and tectonic activity of the time.

The cooling and solidification process was punctuated by periods of extreme bombardment from space debris during what is known as the **Late Heavy Bombardment** (around 4.1 to 3.8 billion years ago), a period in which the Earth was frequently struck by large asteroids and comets. These impacts would have melted much of the early crust repeatedly, creating a hostile environment where any newly formed surface rock was likely destroyed or reabsorbed into the molten mantle. Despite these setbacks, a more stable crust eventually began to emerge.

Early Continental Crust: The Formation of Protocontinents and Island Arcs

The Hadean Eon

As Earth's surface continued to cool, the process of differentiation occurred, wherein lighter materials (like silicates) rose to form the crust, and heavier materials (such as iron and nickel) sank to form Earth's core. This led to the development of different types of crust: **oceanic crust**, which was thinner and denser, and **continental crust**, which was thicker and less dense.

The formation of continental crust was a gradual process that likely began towards the end of the Hadean Eon. Initially, Earth's crust would have been predominantly oceanic, as the planet's surface was dominated by oceans formed from condensed water vapor. However, geological evidence suggests that **small protocontinents** may have existed in the latter part of the Hadean Eon. These protocontinents were not as large or stable as modern continents, but they represented the early seeds of what would later grow into Earth's major landmasses.

It is believed that **island arcs**, chains of volcanic islands formed by subduction processes, played a key role in the formation of early continental crust. As tectonic plates began to move, oceanic plates would have subducted (or slid beneath) other plates, causing volcanic activity and the formation of island arcs. Over time, these island arcs would have accreted (merged) to form larger landmasses.

The composition of the early continental crust was likely granitic, formed from the partial melting of oceanic crust during subduction. This granitic material was more buoyant than the oceanic crust, allowing the protocontinents to persist and gradually expand over time. While these protocontinents were relatively small and geologically unstable compared to modern continents, they were crucial in establishing the foundations for Earth's future landmasses.

The Oldest Rocks: Insights from Earth's Ancient Geological Record

One of the most intriguing aspects of studying the Hadean Eon is the search for the oldest rocks on Earth, which provide direct evidence of the planet's earliest geological history. Although most of the rocks from the Hadean Eon have long since been destroyed by tectonic activity, some ancient remnants have survived.

Among the oldest known rocks on Earth is the **Acasta Gneiss**, located in the Canadian Shield of northwestern Canada. These rocks have been dated to be approximately 4.03 billion years old, making them some of the oldest crustal material on the planet. The Acasta Gneiss represents highly metamorphosed granitic rock, likely derived from the early continental crust. The discovery of

The Hadean Eon

these ancient rocks has provided critical insights into the processes that shaped the early Earth, including the formation of continental crust and the planet's cooling history.

Another significant geological find is the presence of ancient **zircon crystals** in the Jack Hills region of Western Australia. These zircon crystals, which have been dated to around 4.4 billion years old, are among the oldest materials ever found on Earth. While the rocks surrounding the zircons have long since eroded away, these tiny crystals offer a rare glimpse into the conditions of early Earth during the Hadean Eon. The chemical composition of these zircons suggests that liquid water may have existed on Earth's surface as early as 4.4 billion years ago, providing tantalizing evidence that conditions suitable for life could have existed much earlier than previously thought.

Other ancient rock formations, such as the **Isua Greenstone Belt** in Greenland (dating back to about 3.8 billion years ago), provide additional clues to the Earth's early crustal processes. These rocks contain evidence of both volcanic activity and the presence of water, further supporting the idea that Earth's surface was geologically active and dynamic during the latter part of the Hadean.

The study of these ancient rocks has also led to the discovery of early geochemical cycles and the interaction between Earth's surface and atmosphere. For example, some of the oldest rocks show signs of weathering, indicating that rain and other weathering processes were already at work during the Hadean, contributing to the breakdown of rocks and the formation of early soils.

Conclusion: A Dynamic and Formative Era

The formation of Earth's crust during the Hadean Eon was a time of incredible dynamism and complexity. From the cooling of Earth's molten surface to the emergence of the first protocontinents, the processes that shaped the early crust laid the foundation for the planet's long-term geological stability. The study of the oldest rocks, such as the Acasta Gneiss and ancient zircons, has provided crucial insights into the conditions of the Hadean, offering a rare glimpse into a period that shaped the Earth as we know it.

While the exact details of how the first crust formed are still under investigation, the evidence suggests that Earth's early crust was continually recycled and reshaped by volcanic activity, impacts from space, and tectonic processes. Despite the challenges of studying this ancient period, ongoing research continues to reveal new information about the Hadean Eon, shedding light on how Earth transitioned from a molten, hostile environment to a more stable planet capable of supporting life.

Chapter 5: The Early Atmosphere and Oceans

The Hadean Eon represents a time when Earth was still in the throes of its formative chaos, yet it was during this period that the early building blocks of Earth's atmosphere and oceans began to take shape. These processes were fundamental to the eventual emergence of life, as the planet transformed from a molten, inhospitable body into one capable of supporting liquid water and a gaseous envelope.

Formation of the Atmosphere

One of the defining features of the early Earth was its lack of a stable atmosphere. At the time of the planet's formation, the proto-Earth was bombarded by planetesimals and was largely molten due to the immense heat generated from accretion, radioactive decay, and collisions with large celestial bodies. This high-energy environment stripped away any early, thin atmosphere that might have existed. As Earth gradually grew through the accretion of debris from the solar nebula, the process of **volcanic outgassing** became the dominant mechanism responsible for creating the planet's second, more substantial atmosphere.

The Hadean Eon

Volcanic activity on early Earth was extremely intense. The mantle was highly volatile, and frequent eruptions released gases trapped within the planet's interior. This process of degassing released vast quantities of gases that would have formed the first significant atmosphere. The primary gases expelled during this time were **water vapor (H_2O), carbon dioxide (CO_2), nitrogen (N_2), methane (CH_4), hydrogen sulfide (H_2S), ammonia (NH_3)**, and smaller amounts of other volatile compounds. Oxygen was notably absent from the Hadean atmosphere, as free oxygen only became significant billions of years later during the Great Oxygenation Event in the Archean Eon.

This initial atmosphere was highly toxic by modern standards and bore little resemblance to the life-supporting mix of gases we rely on today. It was dominated by carbon dioxide, water vapor, and nitrogen, which combined to create a thick greenhouse effect. This greenhouse effect played a critical role in keeping Earth's surface temperatures high despite the Sun being less luminous in the early stages of its life.

One of the major challenges Earth faced during the Hadean was maintaining a stable atmosphere. The planet was continuously bombarded by asteroids, comets, and other space debris, which occasionally stripped away parts of the atmosphere, particularly during the **Late Heavy Bombardment** period around 4.1 to 3.8 billion years ago.

However, volcanic activity continuously replenished the atmosphere with more gases, ensuring its persistence despite these interruptions.

Emergence of Water: Theories of Water Delivery to Earth

Water is the most critical element for life on Earth, and understanding how water came to exist in liquid form on the planet has been a subject of intense scientific debate. Two main theories have been proposed to explain how Earth's early oceans came into existence.

1. Degassing of the Mantle

The **degassing hypothesis** suggests that water was present in the materials that formed Earth and was released through volcanic outgassing as the planet cooled. According to this theory, as Earth's mantle differentiated and melted, volatile compounds, including water vapor, were released into the atmosphere along with other gases during volcanic eruptions.

Water exists within minerals in the Earth's mantle, and as magma rises and erupts at the surface, water is expelled as a vapor. Over millions of years, these volcanic processes could have released enough water vapor into the atmosphere to eventually condense and form liquid water on the planet's surface.

This theory is supported by the fact that water is found in magmatic processes and is released in modern volcanic eruptions. Furthermore, isotopic studies of Earth's mantle suggest that the ratio of hydrogen isotopes in Earth's water is consistent with what is expected from water that originated within the planet's interior. This points to the possibility that much of Earth's water came from the degassing of its own mantle.

2. Extraterrestrial Delivery: Comets and Asteroids

Another prominent theory proposes that water was delivered to Earth through **cometary and asteroid impacts**. During the Hadean, the young Earth experienced frequent collisions with large bodies, many of which were rich in water or other volatile compounds.

Comets, which are composed of ice, dust, and rocky material, are one of the prime candidates for this delivery. The idea is that comets, bombarding Earth in its early history, brought large quantities of water. Similarly, carbonaceous chondrites, a type of water-bearing asteroid, could have delivered significant amounts of water to Earth's surface. These asteroids contain hydrated minerals, suggesting that they carry water as part of their composition.

Isotopic evidence also supports this hypothesis to some extent. Measurements of the hydrogen-to-deuterium (H/D) ratio in Earth's water are similar to the H/D ratio found in certain types of asteroids. However, the H/D ratios in comets do not always match Earth's water, which has led to debate about how much water comets versus asteroids contributed.

Most likely, Earth's water resulted from a **combination of both processes**—outgassing from volcanic activity and the delivery of water from extraterrestrial sources. Both mechanisms were probably crucial in the planet's accumulation of the vast amounts of water necessary to form its early oceans.

Formation of the First Oceans

Once water vapor had been released into the atmosphere, Earth needed to cool sufficiently for liquid water to condense and collect on the surface, forming the first oceans. During the early Hadean, Earth's surface was far too hot for water to exist in liquid form. However, as the planet radiated heat into space and volcanic activity became less intense, surface temperatures eventually began to drop.

The formation of the Earth's crust was a critical turning point in this process. As molten materials cooled and solidified, Earth's surface transformed from a molten magma ocean to a solid, albeit still geologically unstable,

The Hadean Eon

crust. The appearance of this crust would have created a surface upon which water could collect.

At some point during the Hadean, likely around 4.4 billion years ago, the planet's surface temperatures cooled to below 100°C, the boiling point of water. At this stage, the **water vapor in the atmosphere began to condense into liquid form**, leading to the formation of Earth's first oceans.

As water began to condense and accumulate, the process of **precipitation** took place, with water falling as rain onto the hot, solidifying crust. Initially, this water would have rapidly evaporated back into the atmosphere due to residual heat from Earth's surface. However, as cooling continued, this cycle would have slowed, and permanent bodies of water began to form.

The presence of liquid water likely played a significant role in further cooling the planet. Water has a high heat capacity, meaning it can absorb significant amounts of heat without a dramatic increase in temperature. As the early oceans absorbed heat from the crust and the atmosphere, Earth's surface would have stabilized further, creating more favorable conditions for the emergence of a more stable atmosphere and ocean system.

The formation of Earth's first oceans marked the beginning of a global hydrological cycle, with water evaporating from

The Hadean Eon

the surface, condensing into clouds, and falling back as precipitation. This cycle, fundamental to the planet's climate and weather systems, remains one of Earth's defining features to this day.

The Role of the Oceans in Earth's Evolution

The formation of the oceans was a turning point in Earth's evolution, as it set the stage for many critical processes. Water is essential for many chemical reactions, including those that may have led to the origin of life. Early oceans likely provided a stable environment for chemical compounds to accumulate, react, and perhaps form the first organic molecules.

Moreover, the oceans acted as a carbon sink, absorbing large amounts of carbon dioxide from the atmosphere. This absorption would have moderated the greenhouse effect over time, helping to further cool the planet's surface. The formation of carbonates in the ocean, through the precipitation of dissolved carbon dioxide, played a role in locking away CO_2 in solid form, preventing it from re-entering the atmosphere.

As Earth entered the Archean Eon around 4 billion years ago, the cooling, condensation, and stabilization processes that began in the Hadean laid the groundwork for a more stable climate and the potential for life. The early atmosphere and oceans of the Hadean thus represent the

The Hadean Eon first steps in Earth's transformation from a volatile, molten world into the habitable planet we know today.

The Hadean Eon

Chapter 6: Life in the Hadean?

The Hadean Eon, spanning from approximately 4.6 billion to 4 billion years ago, represents the earliest and most hostile period in Earth's history. Its name, derived from "Hades," reflects the inferno-like conditions on the surface of the young planet, with rampant volcanic activity, frequent cosmic impacts, and a toxic atmosphere. Yet, despite these harsh conditions, scientists have long speculated whether life could have originated during this tumultuous period.

Conditions for Life: Could Life Have Existed in the Extreme Environment of the Hadean?

The surface of Hadean Earth was extraordinarily inhospitable by modern standards. It was dominated by molten rock, magma oceans, and intense volcanic eruptions. The planet was continuously bombarded by asteroids and comets, adding to its violent atmosphere. Temperatures on the surface were likely far too high for life as we know it to survive. Additionally, the early atmosphere was thick with gases like carbon dioxide, methane, ammonia, and water vapor, but devoid of free oxygen, making it unsuitable for aerobic organisms.

However, beneath this hostile surface environment, there may have been refuges where life could have originated.

The Hadean Eon

As Earth gradually cooled, some regions may have formed shallow oceans or pools of liquid water. These bodies of water, combined with volcanic activity and the energy released from hydrothermal vents, could have created environments rich in chemicals that are essential for life. Such environments might have resembled what we now call "extreme environments," like those found around modern-day hydrothermal vents or in sulfur-rich hot springs, where life exists in the form of extremophiles—microorganisms that thrive in conditions considered inhospitable by most other life forms.

Theories on the Origin of Life

Several hypotheses attempt to explain how life could have originated in the Hadean Eon, despite the extreme conditions. The leading theories include abiogenesis, which suggests life arose from non-living matter, and the hypothesis that hydrothermal vents provided the conditions necessary for life to emerge. Let's explore these ideas in more depth.

1. Abiogenesis

Abiogenesis is the idea that life arose naturally from non-living chemicals through a series of chemical reactions. This theory suggests that Earth's early oceans and atmosphere were filled with the right conditions and chemical compounds to produce the first simple molecules necessary for life. Over time, these molecules would have evolved into more complex organic compounds, eventually giving rise to self-replicating molecules, the precursors of modern cells.

One of the most famous experiments supporting this theory is the Miller-Urey experiment (1953), where scientists Stanley Miller and Harold Urey simulated early Earth conditions by mixing water, methane, ammonia, and hydrogen in a closed system and subjecting it to electric sparks to mimic lightning. After a week, they discovered that several amino acids—the building blocks of proteins—had formed. This experiment provided evidence that organic molecules essential to life could have formed in Earth's early environment.

In the context of the Hadean Eon, volcanic activity, frequent lightning, and ultraviolet radiation from the Sun could have provided the energy needed to drive these chemical reactions. As the planet cooled and liquid water became more abundant, the formation of stable, life-sustaining molecules would have become more likely. These molecules might have accumulated in "primordial soup" environments—shallow pools or bodies of water enriched with organic compounds—where they could interact and form increasingly complex structures. Over millions of years, these structures could have evolved into protocells, the ancestors of modern cells.

2. Hydrothermal Vents Hypothesis

Another compelling theory about the origin of life during the Hadean focuses on hydrothermal vents at the bottom of the early oceans. Hydrothermal vents are fissures in Earth's crust where heated water rich in minerals is released into the ocean. These vents create unique ecosystems that are not dependent on sunlight, but rather on chemosynthesis—a process in which microorganisms derive energy from chemicals like hydrogen sulfide.

In modern times, hydrothermal vent ecosystems host extremophiles, organisms that thrive in extreme heat, pressure, and toxic conditions. Scientists have found entire communities of life surrounding these vents, such as tube worms, crabs, and various species of bacteria, living in complete darkness, feeding on chemicals spewed from the Earth's mantle.

During the Hadean, hydrothermal vents could have provided the ideal conditions for life to originate. The combination of heat, water, and a rich supply of minerals might have driven the chemical reactions necessary for the formation of complex organic molecules. These vents also provided a stable environment compared to the chaotic surface conditions. The presence of energy sources like hydrogen sulfide and the protection from cosmic radiation in deep ocean environments may have enabled the earliest forms of life to thrive.

One specific hypothesis is the "alkaline hydrothermal vent theory," which proposes that life began in the naturally occurring proton gradients around alkaline vents. These gradients could have driven the formation of simple organic molecules, which, over time, might have assembled into more complex, self-replicating systems, laying the groundwork for the first primitive forms of life.

3. Panspermia Hypothesis

An alternative, though more speculative, theory is that life did not originate on Earth at all but was instead seeded from elsewhere in the universe. This hypothesis, known as panspermia, suggests that life or its precursors arrived on Earth via meteorites or comets. Given the intense bombardment of Earth by space debris during the Hadean, it is possible that life's building blocks were delivered from space, perhaps even from microbial life forms that survived the journey through space.

While panspermia does not explain how life initially arose, it opens up the possibility that life is a universal phenomenon and that Earth was simply one of many planets to be seeded with life.

Earliest Traces of Life

The Hadean Eon

While it is challenging to find direct evidence of life from the Hadean Eon, scientists have uncovered clues that suggest life might have started earlier than previously thought, possibly toward the end of the Hadean.

1. Isua Supracrustal Belt

The Isua Supracrustal Belt in Greenland contains some of the oldest known rocks, dating back about 3.8 billion years, close to the transition from the Hadean to the Archean. Within these rocks, scientists have found chemical signatures in the form of isotopic ratios of carbon that suggest biological activity. Specifically, these signatures point to a preference for lighter carbon isotopes, a hallmark of biological processes. While the Isua rocks do not contain fossilized cells, the carbon isotope evidence suggests that microbial life may have existed around 3.8 billion years ago, possibly even earlier.

2. Zircon Crystals

Tiny zircon crystals from Western Australia, some of which date back to 4.4 billion years ago, provide another line of indirect evidence for life during the Hadean. Although zircons are not fossils, they preserve isotopic information about the environment in which they formed. Some zircons contain traces of carbon isotopes that hint at the presence of life, suggesting that Earth's surface was cool enough for liquid water and potentially life as early as 4.1 billion years ago.

3. Stromatolites

Stromatolites are layered structures created by the activity of microbial mats, particularly cyanobacteria, and they represent some of the oldest known evidence of life on Earth. The oldest confirmed stromatolites date to around 3.5 billion years ago in the Archean, but there is ongoing debate about whether similar structures could have existed earlier, perhaps even at the end of the Hadean. If stromatolites did exist at that time, they would represent the earliest complex communities of microorganisms on Earth.

Conclusion: The Possibility of Hadean Life

The question of whether life could have existed during the Hadean Eon remains a topic of scientific debate, but emerging evidence suggests that it is not impossible. While the surface of Hadean Earth was hostile, deep ocean environments, particularly around hydrothermal vents, may have provided stable and energy-rich environments where life could have originated. Theories like abiogenesis and the role of hydrothermal vents, coupled with tantalizing isotope data from ancient rocks, suggest that life may have had an incredibly early start on Earth—perhaps even during the last phases of the Hadean Eon.

The Hadean Eon

As technology improves and more ancient samples are studied, our understanding of the Hadean and the origins of life on Earth will continue to evolve. Although direct fossil evidence from this time is scarce, the clues we do have offer a fascinating glimpse into how life might have emerged from the fiery chaos of Earth's earliest eons.

Chapter 7: Earth's Transformation Towards Stability

The Hadean Eon was a time of intense geological, atmospheric, and cosmic activity. Over the span of nearly 600 million years, the Earth transitioned from a molten, inhospitable body to a more stable planet, setting the stage for the development of the solid crust, the formation of oceans, and the emergence of life. However, this transformation was not a simple or rapid process—it involved dramatic shifts in Earth's internal and external systems, including the beginning of tectonic activity, the formation of the planet's magnetic field, and the eventual stabilization of its surface.

Plate Tectonics: Early Tectonic Activity and the Formation of Subduction Zones

One of the most crucial developments in Earth's transformation was the onset of **plate tectonics**. Plate tectonics is the movement of large, rigid sections of Earth's lithosphere (the outer layer) over the underlying, more fluid-like asthenosphere. Today, plate tectonics is responsible for the creation of mountains, earthquakes, and volcanoes, but its origins trace back to the Hadean Eon.

The Hadean Eon

Early Tectonic Activity: A Hypothesis

The Earth's crust during the Hadean was incredibly dynamic, driven by the planet's internal heat, which was much greater than it is today due to the residual heat from planetary formation and the radioactive decay of elements. These extreme conditions caused **convection currents** in the mantle, the layer beneath the crust. These currents likely influenced the movement of early surface materials.

While modern-style plate tectonics as we know it likely did not exist in the Hadean, **proto-tectonics**—a more rudimentary version—may have begun. During this period, Earth's crust was in a constant state of formation, destruction, and reformation. The thin, fragile crust, composed of solidified magma, would frequently crack and shift, resulting in **short-lived plates**. These plates were small, unstable, and may have been constantly recycled back into the mantle through a process similar to **subduction**, where one plate is forced beneath another.

The First Subduction Zones

One of the most important features of tectonic activity is the presence of **subduction zones**. In modern tectonics, subduction zones form when a denser oceanic plate converges with a lighter continental plate and is forced downward into the mantle. This process generates intense geological activity, including earthquakes and volcanic

The Hadean Eon

eruptions. During the Hadean, however, subduction zones may have been short-lived and more chaotic due to the instability of the crust.

The early Earth likely saw the creation of **volcanic arcs**—chains of volcanoes formed at subduction zones where the descending plate melts and causes magma to rise to the surface. These volcanic arcs would have been an important part of the planet's early tectonic regime, contributing to the formation of small landmasses or proto-continents.

Proto-continents were the first large, stable chunks of solid crust that began to form during the late Hadean. While they were much smaller and more fragmented than the continents we see today, their formation marks a critical step toward the development of larger landmasses in the Archean Eon.

The **buoyancy** of early continents also played a role in their stabilization. Continental crust, which is composed of less dense materials such as granite, "floats" on the denser mantle. As early continents formed, they became increasingly stable due to this buoyant nature, allowing them to persist rather than be completely subducted.

Magnetic Field Formation: Earth's Shield Against Solar Radiation

The Hadean Eon

Another key transformation during the Hadean was the possible formation of Earth's **magnetic field**. Today, Earth's magnetic field acts as a shield, protecting the planet from the harmful solar wind—a stream of charged particles emitted by the Sun that can strip away an atmosphere and erode a planet's surface. Without this protective field, Earth's atmosphere and surface would be bombarded by cosmic radiation, making the planet uninhabitable for life as we know it.

The Dynamo Effect

The origin of Earth's magnetic field is linked to the planet's **core**. Earth has a **differentiated structure**, meaning that it consists of layers with different densities. The innermost layer is the **core**, which is composed of iron and nickel. Over time, as Earth cooled, the heavy elements in the core began to solidify, creating a solid inner core and a liquid outer core.

The **dynamo theory** explains how the motion of the liquid outer core generates Earth's magnetic field. The movement of molten iron and nickel in the outer core creates electrical currents, which in turn generate a magnetic field around the planet. This process is known as the **geodynamo**. While it is still debated exactly when this process began, there is evidence to suggest that Earth's magnetic field may have started forming during the Hadean or shortly

The Hadean Eon

afterward, as the planet's internal structure differentiated and the core began its solidification process.

Protection from Solar Winds

The importance of Earth's magnetic field cannot be overstated. Without it, the planet would have been vulnerable to the intense solar wind and radiation from the early Sun, which was much more active than it is today. This radiation could have stripped away much of Earth's early atmosphere and water, making the planet a barren, lifeless rock similar to Mars.

The magnetic field acts as a **magnetosphere**, deflecting most of the solar wind and trapping harmful radiation in areas like the Van Allen belts. This protection allowed Earth's early atmosphere to stabilize and retain essential gases like nitrogen and carbon dioxide, which were critical for the development of future life.

The Transition to the Archean Eon: The End of the Hadean

The final stages of the Hadean marked a crucial turning point in Earth's history. As the planet cooled, the chaotic processes that dominated the Hadean—constant volcanic eruptions, intense bombardment from space, and a semi-molten surface—gradually subsided. This transition into the **Archean Eon** signaled the beginning of a more stable Earth.

The Cooling of the Earth's Surface

One of the most significant developments was the cooling of Earth's surface, allowing for the formation of a **solid crust**. By the end of the Hadean, the surface had cooled enough that a more permanent, solid crust could form over much of the planet. This crust was thicker and more stable than the thin, fragile layers of the earlier Hadean.

This crust may have given rise to the first **proto-continents**, small landmasses that would eventually grow into larger continental plates during the Archean. The stabilization of the crust was crucial for the next stages of Earth's evolution, as it allowed for the retention of surface water and the development of oceans, which played a key role in moderating Earth's temperature and climate.

The Formation of Oceans

As Earth's surface cooled, water vapor in the atmosphere began to condense, leading to the formation of the first **oceans**. These oceans, along with the atmosphere, played a crucial role in regulating Earth's temperature and helped to stabilize the planet's surface. By the time the Archean began, Earth had likely developed a significant amount of surface water, setting the stage for the eventual emergence of life.

The Dawn of the Archean

The **Archean Eon**, which followed the Hadean, is characterized by the further development of Earth's crust, the onset of modern plate tectonics, and, eventually, the emergence of life. The transition from the Hadean to the Archean marks a shift from a chaotic, molten world to a more stable, geologically active planet with oceans, a solid crust, and a magnetic field.

The Hadean laid the foundation for all future Earth processes. Without the intense heat, tectonic activity, and volatile conditions of the Hadean, Earth would not have developed the solid crust, protective atmosphere, and oceans necessary for life. While the Hadean was a time of fiery transformation, it was these very processes that made the planet stable enough to support the complex biological systems that would emerge in the eons to come.

This pivotal chapter in Earth's history highlights the immense forces at play in shaping a planet, transforming it from a violent, molten sphere to a world capable of harboring life. The transition to the Archean marked the beginning of a more recognizable Earth, where life would eventually take hold and flourish.

Chapter 8: Hadean Mysteries and Scientific Exploration

The Hadean Eon, spanning from the formation of Earth around 4.6 billion years ago to roughly 4 billion years ago, remains one of the most elusive and mysterious periods in Earth's history. Although it laid the foundation for everything that followed, studying the Hadean presents unique and profound challenges. Due to the extreme and chaotic conditions of the time, very little direct evidence remains. However, scientists have developed innovative methods to explore this primordial eon, drawing on clues from meteorites, lunar rocks, and ancient minerals. As our technology and techniques advance, the potential for future discoveries continues to grow, offering tantalizing possibilities for unraveling Earth's deepest mysteries.

Challenges of Studying the Hadean: Lack of Surviving Rocks and Direct Evidence

The Hadean Eon is, in many ways, a geologist's nightmare. One of the primary challenges in studying this period is the near-total absence of surviving rocks from that time. The name "Hadean" itself, derived from Hades, the Greek god of the underworld, reflects the hellish conditions that existed during Earth's infancy—molten lava oceans, frequent asteroid impacts, and an atmosphere toxic by

The Hadean Eon

today's standards. These extreme conditions ensured that the early crust of the Earth was continuously recycled, melted, and reshaped by tectonic and volcanic activity.

The Earth's surface during the Hadean was so unstable that almost all the solid material that may have formed has been obliterated. Over billions of years, plate tectonics further eroded the record, constantly subducting and recycling any ancient crust that might have existed. The process of erosion, volcanic activity, and the movement of Earth's tectonic plates means that virtually no rocks from this time have survived to the present day.

Additionally, Earth's atmosphere and hydrosphere were also forming during the Hadean, which likely contributed to the chemical weathering and destruction of early rock formations. As a result, direct evidence from the Hadean period is incredibly sparse, and geologists must rely on rare, indirect clues embedded in later formations.

To date, the oldest known rocks, such as the Acasta Gneiss from Canada, date back to around 4 billion years ago—just at the tail end of the Hadean. These rocks provide only a glimpse of Earth after the most violent epochs had passed. However, even these fragments are tantalizingly incomplete, leaving scientists with more questions than answers about what truly transpired during the Hadean Eon.

The Hadean Eon

Meteorites and Moon Rocks: Evidence of Early Earth Conditions

Given the scarcity of Hadean rocks on Earth, scientists have turned to extraterrestrial sources of evidence, such as meteorites and lunar rocks, to understand the conditions of early Earth. Meteorites, particularly those that date back to the formation of the solar system, act as time capsules from the period when Earth was still forming. These space rocks provide invaluable information about the materials that were present in the early solar system and the processes that led to planet formation.

One of the most significant types of meteorites in this context is the chondrite, which is thought to be composed of some of the oldest materials in the solar system. Chondrites contain small, spherical mineral grains called chondrules that are believed to have formed from the condensation of gas and dust in the solar nebula. By studying the chemical composition of these meteorites, scientists can infer the kinds of materials that were present when Earth was forming, giving us a snapshot of the building blocks that would eventually become our planet.

Another key source of information comes from lunar rocks, brought back to Earth by the Apollo missions. Since the Moon is thought to have formed as a result of a massive impact between Earth and a Mars-sized body (Theia)

59

The Hadean Eon

during the Hadean Eon, lunar rocks offer a unique perspective on Earth's early history. Unlike Earth, the Moon lacks significant geological activity, such as plate tectonics or erosion, meaning that its rocks have remained relatively undisturbed for billions of years.

The lunar highlands, in particular, contain rocks that date back over 4 billion years, offering a window into the conditions of the early solar system and the aftermath of the giant impact that formed the Moon. By studying isotopic ratios in lunar rocks, such as oxygen isotopes, scientists can learn more about the similarities between Earth and the Moon, supporting the giant impact hypothesis and providing insights into the state of the Earth immediately after this catastrophic event.

In addition to providing clues about Earth's formation, lunar samples and meteorites offer evidence of the intense bombardment that characterized the Hadean. The Late Heavy Bombardment, a period when the inner solar system was bombarded with asteroids and comets, is thought to have occurred around 4.1 to 3.8 billion years ago. The cratering record on the Moon, preserved in its ancient surface, suggests that Earth experienced a similar level of bombardment during this time, which would have had profound effects on its surface and atmosphere.

The Hadean Eon

Future Discoveries: Zircons and Ancient Minerals as Geological Timekeepers

Despite the scarcity of surviving Hadean rocks, one mineral has proven to be an invaluable tool in piecing together the story of Earth's earliest eon: zircon. Zircons are incredibly resilient minerals that can survive geological processes that destroy other rocks and minerals. These tiny crystals, often no bigger than a grain of sand, can contain trace amounts of radioactive elements such as uranium, which decay into lead over time. By measuring the ratio of uranium to lead in zircon crystals, scientists can determine their age with remarkable precision.

The discovery of Hadean-aged zircons, particularly from the Jack Hills in Western Australia, has revolutionized our understanding of this ancient time. Some of these zircons have been dated to as old as 4.4 billion years, making them the oldest known materials on Earth. These tiny crystals provide rare and direct evidence of Earth's early crust and the conditions that existed during the Hadean.

By analyzing the isotopic composition of these zircons, scientists have gleaned important clues about the Hadean environment. For example, the presence of certain oxygen isotopes in Hadean zircons suggests that liquid water may have existed on Earth's surface as early as 4.3 billion years ago—much earlier than previously thought. This finding

The Hadean Eon

has profound implications for our understanding of when Earth's oceans may have formed and when conditions suitable for life might have first emerged.

Future discoveries in the field of zircon research hold the potential to uncover even more details about the Hadean Eon. Advances in analytical techniques, such as secondary ion mass spectrometry (SIMS) and atom probe tomography, allow scientists to study the composition of zircons at a near-atomic level, revealing even more about their formation and the conditions of early Earth.

Beyond zircons, ongoing research into other ancient minerals, such as apatite and titanite, may also provide new insights into the Hadean. Additionally, future space missions, such as those targeting asteroids or other planetary bodies, could return samples that contain material from the early solar system, offering further clues about Earth's formation and the processes that shaped the Hadean.

As our understanding of early Earth continues to evolve, so too does the potential for new discoveries. Each piece of evidence, whether from ancient minerals, extraterrestrial rocks, or sophisticated models, brings us closer to answering some of the most fundamental questions about Earth's origins and the conditions that gave rise to life. The Hadean Eon, though largely shrouded in mystery, remains

The Hadean Eon

a critical chapter in the story of our planet—one that scientists are determined to unravel, one discovery at a time.

The Hadean Eon, while largely inaccessible through direct observation, is slowly being revealed through the persistence of scientists and the development of innovative techniques. From the study of meteorites and Moon rocks to the careful analysis of ancient zircons, researchers are steadily piecing together the story of Earth's violent and chaotic beginnings. Each new discovery sheds light on this distant epoch, revealing a world unlike any we can imagine—a world that, despite its hostile conditions, laid the foundation for the planet we live on today. The future holds the promise of even more groundbreaking revelations, as we continue to push the boundaries of what we know about the earliest chapters of Earth's history.

Chapter 9: Conclusion: The Significance of the Hadean

The Hadean Eon, often regarded as Earth's most tumultuous and mysterious epoch, represents a foundational chapter in the planet's history. Though it lacks abundant rock records and direct evidence, the scientific significance of this era cannot be overstated. The Hadean laid the groundwork for Earth's geology, atmosphere, oceans, and even life itself. As the chaotic conditions settled, Earth began its transformation from a molten, inhospitable sphere into a dynamic planet capable of supporting the complex systems we now rely on.

The Hadean as the Foundation of Earth's History

1. Geologic Evolution:

During the Hadean, Earth's initial formation and differentiation into layers (core, mantle, and crust) created the framework for its geologic processes. The Earth's crust, although initially unstable and frequently recycled by the intense heat of the planet's interior, began to solidify toward the end of this eon. The constant bombardment by asteroids, meteors, and planetesimals not only added material to Earth but also influenced the formation of its surface.

The Hadean Eon

The heat generated from these impacts and from radioactive decay drove internal processes, such as the formation of magma oceans. As the planet cooled over millions of years, these magma oceans crystallized, leading to the creation of the first solid rock formations. Although most of these primordial crusts were likely destroyed by continuing impacts and tectonic shifts, some fragments may have survived, offering a glimpse into this distant past through ancient rocks such as the Jack Hills zircons, dated at over 4.4 billion years old.

The significance of these early geologic processes is that they laid the foundation for plate tectonics, one of the most crucial systems governing Earth's geology today. While the exact timing of when plate tectonics began remains debated, the differentiation of the Earth's layers during the Hadean provided the conditions necessary for this process. The movement of tectonic plates would eventually lead to the formation of continents, mountain ranges, volcanic arcs, and ocean basins — all vital features of modern Earth.

2. Atmospheric Formation:

The Hadean atmosphere was radically different from the one we breathe today. Initially, Earth's atmosphere was likely composed of hydrogen and helium, gases captured from the solar nebula. However, this early atmosphere was

The Hadean Eon

stripped away by the young Sun's intense solar winds and constant cosmic impacts. The secondary atmosphere that replaced it was primarily composed of gases released from volcanic outgassing — including carbon dioxide (CO_2), water vapor (H_2O), nitrogen (N_2), methane (CH_4), and ammonia (NH_3).

Although toxic by modern standards, this early atmosphere played a crucial role in creating conditions conducive to future life. Water vapor released during volcanic activity, combined with water delivered by comets and asteroids, likely condensed to form the first oceans as Earth's surface temperatures cooled. Carbon dioxide, a major component of this atmosphere, helped trap heat through the greenhouse effect, ensuring Earth did not become a frozen wasteland despite the faint young Sun.

Over time, the gradual accumulation of water and gases in the atmosphere allowed the Earth to transition from a barren, volcanic planet to one with the potential to sustain more stable, life-supporting conditions. This atmosphere would also become the foundation for later processes such as photosynthesis, which would radically alter Earth's chemical makeup during the Archean Eon.

3. Formation of the Oceans:

One of the most significant legacies of the Hadean is the formation of Earth's oceans. Water, whether delivered by cometary impacts or degassed from Earth's interior, was a key component in cooling the planet and stabilizing its surface. The oceans, which began forming toward the end of the Hadean, were critical for regulating Earth's temperature, absorbing carbon dioxide, and facilitating early chemical reactions that may have led to the emergence of life.

The presence of liquid water oceans as early as 4.4 billion years ago is suggested by ancient zircon crystals, which indicate that Earth's surface temperatures were low enough to support water. This early water may have provided a habitat for primitive life, protected from the harsh surface conditions by the depths of the ocean. In addition, the constant mixing of ocean water with volcanic gases and minerals from the crust may have created the ideal conditions for the chemistry of life to develop.

4. The Origins of Life:

Although the Hadean Eon was marked by extreme heat, constant bombardment, and frequent volcanic activity, it also provided the essential ingredients for life. By the end of this eon, Earth's surface had cooled enough to maintain liquid water and stabilize its atmosphere. The building blocks of life, such as carbon-based molecules, likely formed during this time, driven by interactions between water, atmospheric gases, and energy sources like ultraviolet light or hydrothermal activity.

Theories about the origin of life during or shortly after the Hadean include the idea that life may have emerged in hydrothermal vent systems on the ocean floor. These vents, rich in heat and minerals, could have provided the energy and materials necessary for the first simple life forms to develop. Alternatively, life may have originated in shallow pools near the surface, where the combination of water, sunlight, and volcanic activity created favorable conditions.

While no direct evidence of life from the Hadean has been found, traces of organic molecules and isotopic signatures in ancient rocks from the Archean suggest that life was present by the time the Hadean ended. Thus, the Hadean was not just a time of fiery chaos but also a period that set the stage for one of the most profound events in Earth's history — the emergence of life.

Lessons from the Hadean

The Hadean Eon holds vital lessons not only for understanding Earth's early evolution but also for comprehending planetary formation and evolution across the universe.

1. Planetary Formation in the Solar System:

The Hadean Eon

The Hadean demonstrates that planetary formation is a violent, chaotic process that involves both constructive and destructive forces. Earth's history during this time mirrors the processes that likely occurred on other rocky planets in our solar system, such as Mars, Venus, and Mercury. For example, Mars also experienced a period of intense bombardment, volcanic activity, and possibly the presence of water. However, Mars' smaller size and weaker gravitational field allowed it to cool faster and lose its atmosphere more quickly than Earth, leading to its current barren state.

By studying Earth's Hadean, scientists can draw parallels with other planets and moons, helping them to understand the factors that lead to habitable conditions. The discovery of liquid water and ancient crust on Mars, along with the volcanic activity on moons such as Europa and Enceladus, suggests that similar processes could have occurred throughout the solar system.

2. Exoplanets and the Search for Life:

The Hadean Eon also offers insights into the conditions necessary for life on other planets. The transition from an inhospitable, molten planet to one capable of supporting life suggests that even extreme environments can eventually stabilize to allow the emergence of life. This has implications for the search for life beyond Earth, particularly on exoplanets orbiting distant stars. Planets that experience intense heat and bombardment during their early history may still evolve into habitable worlds if they have the right conditions, such as liquid water, an atmosphere, and geologic activity.

As scientists search for Earth-like planets in other solar systems, understanding the Hadean provides a model for how rocky planets develop and what key markers to look for when evaluating their habitability.

Conclusion: Earth's Fiery Origins and a Blueprint for the Universe

The Hadean Eon, with its intense heat, cosmic impacts, and the birth of oceans and atmosphere, represents the dawn of planetary evolution. It laid the foundation for everything that followed: tectonic activity, climate regulation, the formation of continents, and ultimately, life itself. By studying this mysterious eon, scientists unlock the secrets of planetary formation, providing a blueprint not only for Earth's development but also for understanding the evolution of other planets and the potential for life in the universe.

The Hadean may have been a time of chaos, but it was also a time of creation — the era when Earth transformed from a molten ball of rock into a planet with the potential to harbor life. Understanding this early chapter in Earth's

The Hadean Eon history is crucial for appreciating the dynamic forces that continue to shape our world and for unlocking the mysteries of planetary evolution across the cosmos.

Glossary of Terms

1. **Abiogenesis** – The natural process by which life arises from non-living matter, such as simple organic compounds. It is often considered a possibility for the origin of life during the Hadean Eon.
2. **Accretion** – The process by which dust, gas, and other cosmic materials gradually clump together through gravitational attraction, forming larger celestial bodies like planets, including early Earth.
3. **Acasta Gneiss** – One of the oldest known rock formations on Earth, found in Canada, dating back approximately 4 billion years, providing evidence of early Earth's crust.
4. **Archean Eon** – The geologic eon following the Hadean, lasting from around 4 billion to 2.5 billion years ago, marked by the formation of stable continental crust and the first evidence of life.
5. **Asteroids** – Small rocky bodies orbiting the Sun, remnants from the early solar system's formation. Many of them bombarded early Earth during the Hadean.
6. **Atmospheric Outgassing** – The process by which gases are released from a planet's interior, typically through volcanic activity. This process helped form Earth's early atmosphere.

The Hadean Eon

7. **Basalt** – A type of volcanic rock that forms from the rapid cooling of basaltic lava. It is believed that Earth's first crust was largely basaltic in nature.
8. **Bombardment (Late Heavy Bombardment)** – A period of intense asteroid and comet impacts on the inner planets of the solar system, including Earth, during the Hadean, around 4.1 to 3.8 billion years ago.
9. **Carbonaceous Chondrites** – A class of meteorites that contain water and organic compounds. These meteorites may have contributed to Earth's early water supply during the Hadean.
10. **Core (Earth's Core)** – The innermost part of Earth, composed primarily of iron and nickel. Earth's core began forming during the Hadean as heavier elements sank to the center.
11. **Crust (Earth's Crust)** – The outermost layer of Earth. During the Hadean, the first solid crust began to form as the planet cooled.
12. **Differentiation** – The process by which Earth's materials separated into different layers (core, mantle, and crust) based on density, occurring during the Hadean Eon as the planet cooled.
13. **Early Atmosphere** – The initial gaseous envelope surrounding early Earth, mainly composed of carbon dioxide, nitrogen, methane, and water vapor, with little to no oxygen.

The Hadean Eon

14. **Earth's Magnetic Field** – A magnetic field generated by the motion of molten iron in Earth's outer core. It may have first formed during the Hadean, protecting the planet from solar winds.
15. **Feldspar** – A group of rock-forming minerals common in Earth's crust. Early feldspar-containing rocks may have formed as Earth's surface cooled during the Hadean.
16. **Giant Impact Hypothesis** – The theory that the Moon formed from the debris of a massive collision between Earth and a Mars-sized body (Theia) during the Hadean Eon.
17. **Granite** – A common type of intrusive, igneous rock. Small pockets of granite may have started forming by the end of the Hadean as Earth's crust began stabilizing.
18. **Hadean Eon** – The first eon in Earth's history, lasting from the planet's formation around 4.6 billion years ago to about 4 billion years ago. It is characterized by extreme heat, volcanic activity, and constant meteor impacts.
19. **Heavy Elements** – Elements like iron and nickel, which are denser and sank to form Earth's core during the Hadean Eon as part of planetary differentiation.
20. **Hydrosphere** – The collective mass of water on Earth, including oceans, rivers, lakes, and

The Hadean Eon

underground water. Earth's early hydrosphere began forming during the Hadean as the planet cooled.

21. **Isotope Dating** – A method used by scientists to determine the age of rocks and minerals by measuring the decay rates of isotopes, such as uranium-lead dating used for Hadean rocks.
22. **Late Heavy Bombardment** – A period of frequent collisions between Earth and celestial objects (comets, asteroids) roughly 4.1 to 3.8 billion years ago, thought to have shaped Earth's surface.
23. **Lithosphere** – The rigid outer layer of Earth, consisting of the crust and the upper part of the mantle. During the Hadean, this layer was in the process of forming as the planet cooled.
24. **Magma Ocean** – A theoretical period during which Earth's surface was mostly or entirely molten due to high internal temperatures during the early Hadean.
25. **Mantle** – The layer of Earth located between the crust and the core. The mantle played a key role in volcanic activity during the Hadean, releasing gases that helped form the early atmosphere.
26. **Meteorites** – Fragments of asteroids or comets that survive passage through Earth's atmosphere and impact the surface. Meteorite impacts were frequent in the Hadean and may have brought essential compounds, including water.

The Hadean Eon

27. **Moon Formation** – The process by which the Moon formed, thought to be a result of the giant impact hypothesis, where a collision between Earth and Theia created a debris cloud that coalesced into the Moon.
28. **Nucleosynthesis** – The process that forms new atomic nuclei from pre-existing protons and neutrons. The Hadean saw elements heavier than hydrogen and helium on Earth, thanks to nucleosynthesis in earlier stars.
29. **Oceans (Hadean Oceans)** – Large bodies of water that formed during the Hadean Eon as Earth's surface cooled, and water vapor in the atmosphere condensed.
30. **Outgassing** – The release of gases from volcanic activity. During the Hadean, outgassing helped form Earth's early atmosphere and contributed water vapor that eventually led to ocean formation.
31. **Planetary Differentiation** – The process by which a planet separates into distinct layers (core, mantle, and crust) based on the density of its materials. This process occurred in Earth during the Hadean Eon.
32. **Planetesimal** – Small celestial bodies that formed through the process of accretion during the early solar system. These planetesimals collided and merged to form the planets, including Earth.

The Hadean Eon

33. **Primordial Crust** – The first solid crust that formed on Earth's surface as the molten rock cooled. This crust was unstable and frequently re-melted by ongoing volcanic activity and impacts.
34. **Proto-Earth** – The term used to describe Earth during its earliest formation, when it was still growing by accreting planetesimals and enduring frequent impacts.
35. **Radiometric Dating** – A technique used to determine the age of rocks by measuring the decay of radioactive isotopes. Zircon crystals have been dated to the Hadean Eon using this method.
36. **Solar Nebula** – The rotating disk of gas and dust from which the Sun and the planets formed around 4.6 billion years ago, initiating the events that led to the Hadean Eon.
37. **Solar Wind** – A stream of charged particles released from the Sun's upper atmosphere. During the Hadean, the lack of a protective magnetic field may have allowed solar wind to strip Earth of its early atmosphere.
38. **Subduction** – A process where one tectonic plate moves under another and sinks into the mantle. Though subduction began in later eons, early tectonic activity during the Hadean laid the groundwork for this process.

The Hadean Eon

39. **Tectonic Activity** – Movements of Earth's lithosphere due to internal heat and convection in the mantle. Early forms of tectonic activity began during the Hadean as the planet's surface cooled.
40. **Theia** – A hypothetical Mars-sized body that is believed to have collided with early Earth, leading to the formation of the Moon.
41. **Volcanic Activity** – Intense volcanic eruptions that were common during the Hadean, contributing to outgassing and the formation of Earth's early atmosphere.
42. **Water Delivery (Hadean)** – The process by which Earth's water may have been delivered via comets and water-rich asteroids during the Hadean Eon, contributing to the formation of oceans.
43. **Zircon** – A mineral that forms in igneous rocks and can survive geologic processes for billions of years. Zircons from the Hadean provide some of the oldest evidence of Earth's crust.
44. **Zircon Dating** – The process of using the radioactive decay of uranium within zircon crystals to determine their age. Zircon crystals from the Hadean Eon are among the oldest materials on Earth.

Bibliography

1. "Earth's Deep History: How It Was Discovered and Why It Matters" by Martin J.S. Rudwick

Offers a historical perspective on the scientific discoveries that helped us understand Earth's deep past, including its fiery beginnings.

2. "The Goldilocks Planet: The 4 Billion Year Story of Earth's Climate" by Jan Zalasiewicz and Mark Williams

Examines Earth's climate evolution, beginning with the early Hadean, and explores how climate has shaped the planet over billions of years.

3. "Origins: How Earth's History Shaped Human History" by Lewis Dartnell

This book provides a broader look at Earth's geological history, tracing back to the Hadean eon and explaining how early planetary processes laid the groundwork for life.

4. "A Brief History of Earth: Four Billion Years in Eight Chapters" by Andrew H. Knoll

This book offers a concise overview of Earth's history, starting from the Hadean Eon, making it an excellent resource for understanding the planet's early geological conditions.

5. "The Planet Factory: Exoplanets and the Search for a Second Earth" by Elizabeth Tasker

While primarily focused on exoplanets, this book provides context on planetary formation, drawing parallels to Earth's own fiery origins.

6. "The Early Earth: Accretion and Differentiation" edited by James Badro and Michael J. Walter

A collection of scientific essays that dive into the processes of accretion, differentiation, and early planetary formation relevant to the Hadean.

7. "Cataclysms: A New Geology for the Twenty-First Century" by Michael R. Rampino

Explores catastrophic events in Earth's history, including those that occurred during the Hadean, providing insight into how these shaped the planet's evolution.

8. "Earth: An Intimate History" by Richard Fortey

A comprehensive narrative of Earth's geological evolution, with sections exploring the formation of Earth's crust and early history, including the Hadean Eon.

The Hadean Eon

9. **"Meteorites and the Early Solar System II"** edited by Dante S. Lauretta and Harry Y. McSween Jr.

While focused on meteorites, this book provides valuable information on planetary accretion and impacts, which were crucial during the Hadean Eon

Acknowledgments

The journey of writing *The Hadean Eon: Earth's Fiery Origins and the Birth of a Planet* has been as thrilling as the subject itself, and I owe a great deal of gratitude to the many people who made this book possible.

First and foremost, I would like to express my deepest appreciation to the scientific community—geologists, astronomers, and planetary scientists—whose groundbreaking research and discoveries have provided the foundation for this work. Without their tireless dedication to uncovering the mysteries of Earth's ancient past, this book would not have been possible.

To my family and friends, thank you for your unwavering support and encouragement throughout this writing process. Your belief in my passion for science and storytelling gave me the motivation to dive deep into the Hadean Eon and bring this remarkable period of Earth's history to life.

A special thanks goes to my editor, whose keen eye for detail and invaluable insights helped shape this book into its final form. Your guidance has been instrumental in ensuring that complex scientific concepts are presented in a way that is accessible and engaging for readers.

I would also like to extend my gratitude to the many researchers and experts who kindly answered my questions, pointed me to resources, or inspired me with their work. Your generosity with your time and knowledge has been greatly appreciated.

Finally, to my readers: Thank you for your curiosity and enthusiasm for Earth's earliest chapter. It is my hope that this book ignites your imagination and deepens your understanding of our planet's fiery origins.

This book is a testament to the wonders of science and the endless fascination of exploring our past, and I am grateful to all who have helped me bring it into the world.

With sincere appreciation,
Zahid Ameer
Versatile Indie Author

Disclaimer:

The information presented in *The Hadean Eon: Earth's Fiery Origins and the Birth of a Planet* is intended for educational and informational purposes only. While every effort has been made to ensure the accuracy and reliability of the content, scientific understanding of the Hadean Eon and Earth's early history is continually evolving as new research and discoveries emerge. The theories, data, and interpretations discussed in this book reflect current knowledge and may be subject to change.

This book is not intended to serve as a definitive or exhaustive source on the subject. Readers are encouraged to consult additional scholarly sources and scientific literature for more detailed information. The author and publisher assume no responsibility or liability for any errors, omissions, or interpretations made by readers based on the content provided in this book. All geological, scientific, and historical discussions are intended to foster curiosity and understanding, but should not be taken as absolute or final conclusions.

About me

I am Zahid Ameer, hailing from the vibrant country of India. As an author, ghostwriter, bibliophile, online affiliate marketer, blogger, YouTuber, graphic designer, and animal lover, I have woven my passions into a unique tapestry that defines my life's work.

Born and raised in India, I have always possessed a deep love for literature. With an insatiable appetite for books, I have amassed an impressive collection of around 1,600 titles, predominantly in English. My passion for reading brings me immense joy and serves as a source of inspiration for my writing endeavors.

I have compiled an impressive portfolio of written works as an author and ghostwriter. With a captivating writing style and an innate ability to craft engaging narratives, I bring my stories to life, captivating readers from all walks of life. My wide range of interests and experiences contribute to the richness of my writing, allowing me to connect with my audience on a heartfelt level effortlessly.

Beyond my literary pursuits, I have also established a strong presence on various digital platforms. I utilize my YouTube channel and blog to raise awareness about all types of knowledge and to share heartwarming stories of animals. Using my platform to shed light on important issues, I strive to create a world where humans and animals can coexist harmoniously.

In addition to my work as an author, I have also dabbled in the world of affiliate marketing. With my webpreneur spirit, I have ventured into online marketing, leveraging my knowledge and skills to promote products and services that align with my values.

However, my most cherished role is that of a father. Family is at the core of my being, and everything I do is centered around creating a better future for my loved ones. My dedication to my family is evident in my passion for personal growth and my relentless pursuit of success. Through my various endeavors, I strive to set an example of perseverance and ambition for my children, inspiring them to chase their dreams unapologetically.

In a world where specialization often dominates, I defy convention by embracing multiple passions and excelling in diverse fields. My love for books, animals, and family has become the driving force behind my achievements. By the grace of Almighty God, my unique blend of characteristics has allowed me to leave an indelible mark on the world, enriching the lives of those I encounter along the way.

To your grand success in life,

The Hadean Eon

Zahid Ameer
Versatile Indie Author

www.ingramcontent.com/pod-product-compliance
Lightning Source LLC
Chambersburg PA
CBHW071106240526
45469CB00006BD/2347